CONTENTS

Author's Note 5
History I: From Beams to Beranger 7
Evolution of the Knife Edge 20
History II: The Industry of Precision 21
Examples of Early 19th Century Scientific Weighing Devices 30
Types of Scales and Terms Used 32
Examples of Beam Balances 38
Examples of Steelyards 49
Examples of Spring Balances 53
Examples of Counter Scales 58
Examples of Coin Balances 64
Examples of Letter Balances 67
Evolution of the Pivot 72
Weights and Measures 73
Directory of Marques, Inventors and Manufacturers 81
Pyknometers and Hydrometers 90
Exhibits at the City of Birmingham Museum 93
Acknowledgments 95
Bibliography 95

4

1   A primitive cord-pivot beam scale shown by Caravaggio in the painting of Camillus refusing his consent to the terms of the Gauls, c. 381 BC. The picture is reproduced from a copy of an engraving by Saenredam in the Avery Historical Museum.

VETERAN SCALES
AND BALANCES

Brian Jewell

MIDAS BOOKS

By the same author
  Model Car Collecting
  Veteran Sewing Machines
  Fairs and Revels
  Sports and Games—history and origins

In the Midas Collectors' Library series
  Smoothing Irons
  Veteran Talking Machines
In preparation:
  Motor Badges and Figureheads
  Royal Commemorative China
  Dolls House Furniture
  Tea and Coffee Pots
  Treen and Earthenware
  Toy Theatres and Puppets
  Military Commemorative China
  Sewing Boxes and Paraphenalia
  Early Wireless
  Early Typewriters

First published 1978 by
MIDAS BOOKS,
12 Dene Way, Speldhurst,
Tunbridge Wells, Kent, TN3 0NX

© Brian Jewell, 1978

ISBN 0 85936 081 4

All rights reserved. No part of this publication may be reproduced, stored in a retrieval system, or transmitted, in any form or by any means, electronic, mechanical, photocopying, recording or otherwise, without prior permission of Midas Books.

Printed in Great Britain by Chapel River Press, Andover, Hampshire.

INTRODUCTION

There is something about scales that calls for respect. Standing high above the Central Criminal Court in the Old Bailey, London, is F. W. Pomeroy's figure of Justice (1905), with a beam balance in her hand. One gets the idea that if a lack of respect is shown, some great sword of retribution will descend with dire results.

Civilisation owes existence to long-forgotten thinkers and inventors. Ethics, morality, social administration, are rules by which we all live, whether or not we subscribe wholeheartedly to them; standards which are rooted in pre-history.

As current philosophy is shaped by pre-historic conceptions, so too is our technology. The basic bricks of invention are: the ability to make fire, the wheel, animal harness, the employment of metals, and the facility of measurement. No subsequent discovery has had greater importance than these five.

Technology needs standards of measurement as society needs its rules. The balance gave part of this facility.

Which peoples first used the balance, and when, is a matter for conjecture, but there is some archaeological evidence that it may have been in Asia Minor that weighing devices were conceived.

Scales and balances make ideal collecting subjects. Some with intricate, precision mechanisms, others crude and basic, but all with the stamp of their functional purpose and with individual styles of their makers.

It is surprising that more has not been written on the subject of weighing, and those of us who are fortunate enough to have a copy, have had to rely for much of our guidance on the concise *Short History of Weighing* by L. Sanders—Curator of the Avery Historical Museum (published by W. & T. Avery, 1947), now regrettably long out of print.

I trust *Scales and Balances* will help collectors and others interested in the subject in their quest to find out more about machines in their possession.

BJ

Tunbridge Wells, 1977

2   The reverse of a Roman coin in the Avery Historical Museum depicting Justice, holding in one hand a balance, and in the other a 'cornucopia', or horn of plenty.

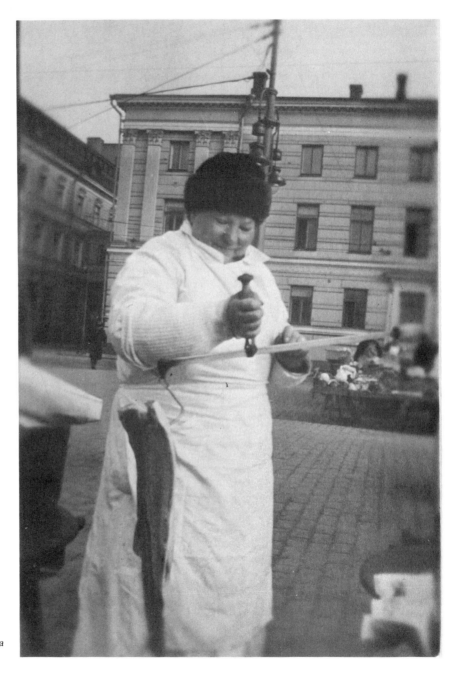

3   Weighing fish in Helsinki, Finland, with a bismar, 1939.

HISTORY I: FROM BEAMS TO BERANGER

The similarity between the beam balance and the load-carrying yoke is more than coincidental. They both depend on the principle of equal masses hung from the ends of a beam in equilibrium, if the beam is supported at the centre. Which came first is purely academic. What is certain is that the yoke was the first of the load carrying aids and the balance was the earliest of weighing devices.

In its simplest form the beam balance is a straight piece of timber pierced with three holes: one near each end and a third in the middle, through which cords or leather thongs are threaded. The balance is suspended from the centre line and the loads hang from the end lines—the standard weight from one end and the load to be compared at the other.

Although adequate for rough and ready assessment of such commodities as wheat, these crude balances, used before 4000 BC, lacked precision. It was difficult to locate the holes, and the cords or thongs moved about in the holes.

A substantial step forward came when some enterprising ancient Egyptian found a way of bringing the cords from the ends of the beam.

The suspended cord-pivot balance is still in limited use to this day in some parts of the world, notably Asia.

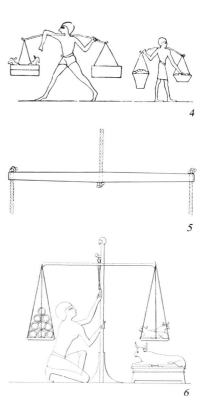

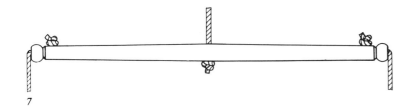

SECTION OF BEAM END.

4   The Egyptian yoke.

5   Primitive cord-pivot beam.

6   Egyptian lotus-ended beam.

7   Indian equal-armed beam, 18–19th century.

History I:
From Beams to Beranger

When the Greeks and Romans adopted the beam balance they usually used bronze as material for the beam. Instead of suspending the beam by a cord or thong, they preferred to use a ring as a pivot. This was hardly an advance on the Egyptian design as the ring tended to wander in the hole, thus varying the length of the arms of the beam.

The principle was long lived and beam balances, either suspended or in supported form, were the commonest weighing devices in Europe until about the beginning of the 16th century.

A variation of the balance was the *bismar* or *auncell*—a lever with a fixed counterweight at one end and a hook at the other on which the load to be weighed is suspended. The fulcrum is a loose loop of cord or leather through which the lever is passed until a point of balance is obtained. The weight is read off the arm at the point of balance; graduations being marked by nail heads in the case of early bismars.

It has been said that the bismar was carried to many parts of Europe and Asia by the nomadic Aryan tribes. Be that as it may, it is fairly certain that the Danes brought the bismar to Britain in the 8th century, and that it was widely used until banned by Edward III, in Statutes between 1351 and 1353.

8   *Esthonian bismar.*

History I:
From Beams to
Beranger

9 A Roman balance preserved at the Avery Historical Museum.

History I:
From Beams to Beranger

The bismar or auncell continued in use in many parts of the world and still survives in the East.

The bismar depends on a sliding fulcrum for its operation whereas the steelyard has a fixed fulcrum near the end of the lever where the load is hung. Sliding along the arm of the lever is a poise to counter-balance the load. When the point of balance is found the weight is read off the graduations marked on the arm.

This type of balance probably came into use a little later than the bismar in Campania on the west coast of what is now Italy, and in time became adopted by the Romans—this is one of the most important developments in the history of weighing, the principles of which are still used today.

Steelyards have been produced in all sizes and for all purposes from those of over 20ft in length for weighing carts to minute examples for checking the weight of coins.

A great advance came in the 16th century when the true knife-edge was evolved. Accuracy had long been a problem for scalemakers. Pivots, which were formed by hooks, rods or cords through holes, suffered from being inaccurate through the wandering of pivot posts or because of friction between relatively large surfaces of the parts. As the need for more precise measurement of weight increased, particularly in the field of chemistry, scale makers faced the challenge and the knife-edge was conceived.

10

11

10 A Roman steelyard in the British Museum.

11 *Justice* by Sir Joshua Reynolds (1723–92). In obedience to the obsession of the time for classical themes, the artist used a steelyard instead of an equal-armed beam as a symbol. The original of the painting in New College, Oxford, of which Sir Joshua was an honorary Doctor of Laws.

12 *Water Mill*, by Phillip Galle (1537–1612). The two steelyards are modern in type with fixed pin pivots; each instrument has two fulcra, for light and heavy loads. It seems that at this period Netherlands farmers carried their own steelyards with them when taking sacks of corn to the miller.

Leonardo da Vinci (1454–1519), one of the most remarkable and many-sided intellects of the Middle Ages, appropriately named 'the Faust of the Renaissance', designed the first recorded self-indicating scale. In fact, his notebooks show two designs, one with a semi-circular chart and the other triangular, although the principle is the same: when an object to be weighed is placed in a suspended pan, the chart acts as a pendulum and finds a new position of equilibrium, the weight being shown on the chart where a plumb bob crosses the face.

Like so many of Leonardo's conceptions, the self-indicating scale was ahead of its time and no practical application of the design was made.

Leonardo's self-indicating scale had to wait until the 19th century before being manufactured. So, too, did the principle devised in 1669 by Gilles Personne de Roberval.

Actually, Roberval's system started as a demonstration model to confound the mathematicians of the time. The model was called the 'Static Enigma' and comprised a beam, legs and stay forming a parallelogram, centrally supported by a stand. From the legs projected two arms on which hung poises of equal weight. What surprised the mathematicians was that balance is maintained even when one poise is moved inwards and the other outwards.

History I:
From Beams to Beranger

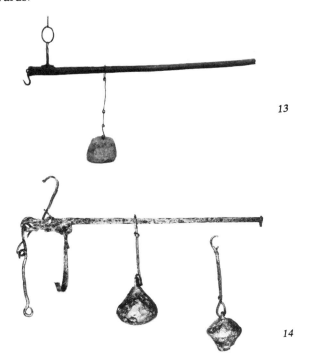

13 Pundler, or wooden steelyard, from Orkney. It bears the stamp of George III and three older obliterated marks.

14 Roman-British steelyard, found at Clipsham, Rutland.

History I: From Beams to Beranger

It did not occur to anyone that this was the basis for a scale design. But indeed it was, and from the beginning of the 19th century many kinds of counter scales were built around the principle, and it is still widely used today.

An impetus to the development of scales came in 1741, in the guise of the Turnpike Act, which gave assent to the trustees of roads to build at their toll gates a 'crane, machine or engine, which they shall judge proper for the weighing of carts, waggons and other carriages'. Tolls had to be charged according to weight and the money so raised used for the maintenance of the roads.

Of course, the Turnpike Act meant increased business for the scalemakers and enormous steelyards were installed. There also had to be a crane so that the vehicle could be hoisted off the ground before its weight could be taken by the scale. This was heavy and time consuming work and it was not long before inventors started putting their minds to the production of more convenient weighing devices.

The answer lay in platform weighing machines, or weighbridges, over which a cart could be driven and the weight ascertained. Several inventors concerned themselves with weighbridge design, among them Eayre, Yeomans and John Wyatt, a mechanic employed by Matthew Boulton of the Soho Manufactury in Birmingham. Wyatt had already invented a spinning mule before turning his attention to weighbridges.

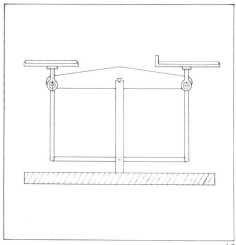

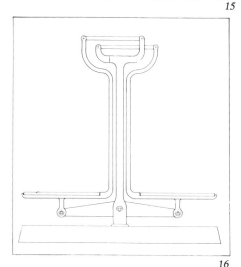

15 Roberval pattern counter scale.

16 Inverted Roberval or Imperial counter scale.

17 Roberval's static enigma

History I:
From Beams to Beranger

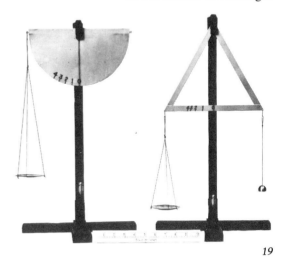

18 Leonardo da Vinci (1452–1519). Self-portrait.

19 Models built by the Research Department of W. & T. Avery Limited, from sketches by Leonardo da Vinci. From his description it is clear that he was aware of the advantages of visible automatic weighing. The models are rather smaller than 'half-an-arm's length' in diameter of dial—specified by Leonardo.

History I:
From Beams to Beranger

20 John Wyatt's weighbridge. From a model of the Lichfield machine (c.1750). The model is now in the City of Birmingham Museum.

History I:
From Beams to Beranger

IT IS PROPER TO POINT OUT THAT THOUGH THE EARLIEST ILLUSTRATIONS OF WEIGHBRIDGES OF THE WYATT TYPE NOW EXTANT SHOW CONICAL PIVOTS IN THE BOTTOM LEVERS IT IS PROBABLE THAT WYATT HIMSELF USED KNIFE-EDGE PIVOTS IN HIS FIRST MACHINES.

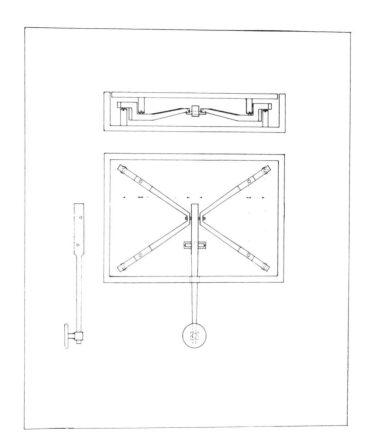

21 John Wyatt's weighbridge. Conical pivots were used for the fulcra and load centres of the two triangular bottom levers. The third lever (transfer lever) is of the first order, and the counterbalancing end is provided with a rigidly attached table for the deposition of the eke-weights (proportional weights), by which the load could be balanced. Hence these old weighing platforms were often referred to as 'bob-up' machines.

History I:
From Beams to Beranger

Wyatt's platform weighing machines, the first of which was installed at the Birmingham Workhouse, comprised a number of levers supporting the platform staging and arranged in such a way that, no matter the position of the load on the platform, the leverage was constant. This is the basis for most of the platform weighing machines which have been produced.

Like so many inventors who have given the world great technical advances, Wyatt was always forced to fight against poverty.

Spring balances began to appear in numbers around 1760, although it is believed that such devices were being made in the previous century. As a piece of weighing equipment the spring balance is a convenience inspired compromise. The great advantage is in portability—most spring balances will fit easily into a pocket—but it does lack the accuracy of other scales because of weakening of the spring through use.

Various forms of spring have been used, the commonest being the helical or coil spring, which features in most of the Salter balances which have remained fairly constant in design since first made by Richard Salter in the mid-18th century. But springs can take other forms: about the time Salters started making their coil spring balance, the sector spring resistant was also being made. It comprised a spring steel strip in the form of a letter V on its side, the opening of which had two curved metal strips each fixed at one end passing through a slot in the opposite end of the V. The balance was suspended from a ring fixed to a free end of one of the strips, while at the lower extremity of the other strip there was a hook which held the load.

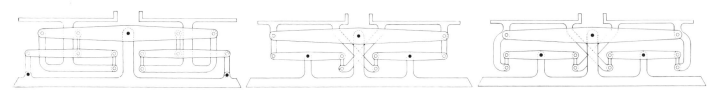

22   *Beranger counter scale and variations.*

History I:
From Beams to Beranger

23 The cart weighing steelyard at Soham.

History I:
From Beams to Beranger

When the ends of the V closed under the weight of the load a reading could be made on a graduated scale on the outer curved strip.

A variation of the sector spring balance had a C shaped spring. One production example of this was called the Mancur balance which became popular in the 19th century.

The early years of that century saw a number of inventors working on various forms of spring balance, among them Augustus Siebe, at first alone and later in partnership with H. Marriott. Together they developed the elliptical spring balance which was manufactured from 1853 by Henry Pooley & Son under an arrangement with Marriott's son.

Pooley used elliptical spring dial mechanisms on platform scales which were built under the Fairbanks' agreement (see page 21).

By this time—the early 19th century—most counter scales were based on the Roberval lever principle which had its origins in the 'Static Enigma' of 1669. there were a number of derivatives including one called the 'Imperial', in which an arrangement of legs and stays were set above the beam, the weight and the platform.

Another Roberval variation was developed by a French scalemaker, Joseph Beranger, in the mid-19th century. Beranger's system was more complicated and expensive to make but it was reliable and accurate. In fact, it still enjoys considerable popularity, particularly in Europe.

SCALES AND STEELYARD. 11th century

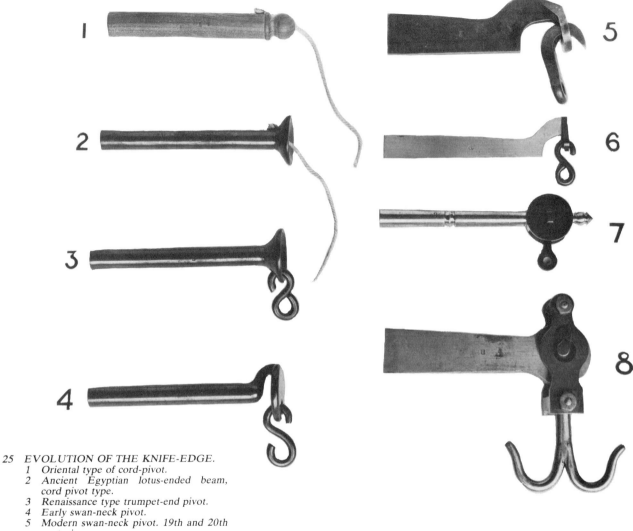

25 EVOLUTION OF THE KNIFE-EDGE.
1. Oriental type of cord-pivot.
2. Ancient Egyptian lotus-ended beam, cord pivot type.
3. Renaissance type trumpet-end pivot.
4. Early swan-neck pivot.
5. Modern swan-neck pivot. 19th and 20th centuries.
6. German swan-neck pivot. c. 1880–1890.
7. Box-end beam pivot. From 16th to end of 19th century.
8. Dutch-end beam pivot. 18th and 19th centuries.

HISTORY II: THE INDUSTRY OF PRECISION

It was in 1831 that the American brothers, Thaddeus and Erastus Fairbanks, were granted a US Patent covering a compound lever platform scale. Two years later an application was made on their behalf for a British patent. This was granted but subsequently withdrawn when it was pointed out that the principle had been used in scale production for some time. The principle, in fact, went back nearly 100 years when John Wyatt built his weighbridges.

The Fairbanks' design, however, did feature some important innovations and scalemaker Henry Pooley saw in it a rich potential. Railways were the new and growing form of transport and it was obvious that numbers of scales for weighing parcels and goods would be needed.

Pooley came to an arrangement with Fairbanks to manufacture these platform scales and delivered one of his first productions to the Manchester-Liverpool Railway in 1835.

The early Pooley platform scales combine the principles of the steelyard with the lever system associated with weighbridges. Apart from a sliding poise to indicate the lower weight range—usually up to one stone—proportional weights were suspended from the end of the steelyard.

About the same time as Pooley was manufacturing scales for railways in Britain, the Quintenz platform scale was coming into use on the continent of Europe. Like the Fairbanks' design and that of Wyatt before them, the Quintenz uses the weighbridge lever system but this time a single lever instead of a multiple arrangement. This acts directly to the steelyard at the rear of the machine.

In time, Pooley was able to dispense with the loose proportional weights and his later productions use a series of poises.

These platform scales are of considerable capacity, capable of weighing several tons. Refinements were introduced, such as devices for automatically printing the weight and fare on a ticket.

The railways certainly accelerated the development of weighing machines and provided a huge demand for the scalemakers' products. So, too, did the growing postal services in the ounce and pound bracket. Never before had there been a need for so many cheap but accurate machines.

In its commonest form the postal balance for household and office is a miniature Roberval counter scale. Gleaming brass examples of these grace every antique shop through the length and breadth of the land, often commanding ridiculously high prices quite out of proportion to their rarity value. They look nice, people who wouldn't know a Pooley from a

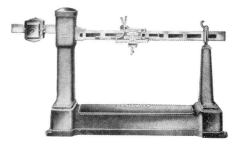

26 Early Pooley platform scale. Mid-19th century.

27 Large ticket-printing steelyard. Early 20th century.

History II:
The Industry of Precision

28 Quintenz platform scale. Introduced in France and Germany c. 1820.

Wilkinson buy them to grace their shelves, the dealers know it and up go the prices accordingly. Such is the agony of serious collectors in a world of fickle fashion.

The need for cheap postal balances exercised the ingenuity of the scalemakers. Apart from the Roberval pattern scales and the various sizes of Salter spring balances, there was made a delightfully simple pendulum balance, patented in 1863 and widely sold, not only for weighing letters but for anglers who wanted a simple trouble-free device to assess their catches at the riverbank.

The pendulum balance in this form comprised a flat piece of brass cut approximately in the shape of a teardrop. The load to be weighed is suspended from the neck of the teardrop. The balance is then held by a pivot which appears at the top of the teardrop when in an inclined position. The bulk of the teardrop now acts as a resistant. The weight is read off from graduations engraved on the lower part of the teardrop, indicated by a free and always perpendicular needle.

It is easy to identify features of the pendulum balance with those of the historic bismar, and with Leonardo's self-indicating scales.

Another self-indicating pendulum resistant scale of the 19th century is a standing balance with the load pan at the top. A lever connects it to a curved plate marked with graduations. At the end of this plate is a resistant weight. The weight of the load is read off the plate, indicated by a static needle.

The pendulum principle has been used for many self-indicating types of scales. In the early years of this form of counter scale (as distinct from the spring balance) the graduations on the chart were of unequal width, increasing with the weight of the load. It was only in 1906 that a means of attaching the connecting rod between the load pan and the pendulum via a cam was found, making it possible for the chart to show equal graduations.

The modern self-indicating counter scale is a combined system of Beranger levers and pendulum resistant.

It was a logical step from an indicator chart, showing the weight, to one that also showed the price of the article being weighed. The 19th century inventors were never slow to accept a challenge and by 1900 several designs of price-indicating weighing machines were available to shopkeepers.

One, the Stimpson, was sold in America from 1897. This scale uses a steelyard with an indicator attached to the poise. This traverses a rectangular chart covering weights up to 10lb, graduated so any price may be read.

The disadvantage of the Stimpson, of course, was that the poise had to be moved manually which made its operation very slow. Within a short while this defect was rectified by the Dayton Computing Scale Company of Dayton, Ohio. Their product was an automatic machine with a hanging load pan, balanced by two helical springs, the cylindrical chart being rotated by a rack and pinion mechanism.

History II:
The Industry of Precision

29 Quadrant letter balance. Late 19th century.

30 Typical mechanism for self-indicating counter scales using a pendulum for the resistant unit. Load is applied to the pendulum by means of a steel band attached to a cam. The steel band is connected to the lever system supporting the scale pan. Introduced c. 1906.

History II:
The Industry of Precision

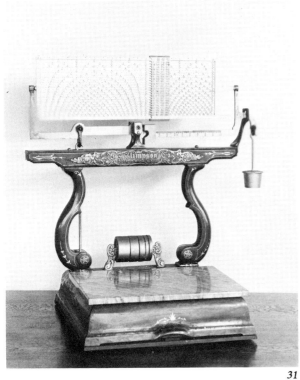

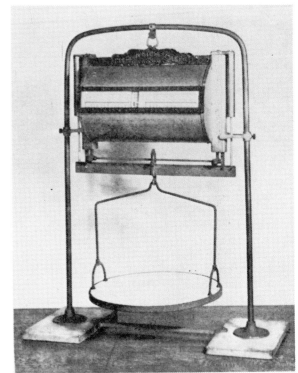

31 Stimpson steelyard computing scale.

32 Dayton computing scale with suspended pan. Late 19th century.

Dayton's second design had the load pan supported by a Roberval lever system, and it was this model that was produced and marketed in Britain from 1907 by Henry Pooley & Son Limited under licence, and given the name 'Royal'.

Another American manufacturer making price-indicating scales was the Toledo Scale Company. Their product used a pendulum resistant patented by De Vilbiss, in place of helical springs of the Dayton machine. It was the Toledo that was marketed in England by W. & T. Avery Limited in competition with the Pooley-made Dayton 'Royal'.

Up to now we have been considering self-indicating and computing scales suitable for shop use. By their design they were limited to weighing comparatively light loads. For heavy industrial purposes other self-indicating systems had to be evolved.

Henry Pooley & Sons Limited introduced a hydrostatic balance in 1874. This was a beam type of balance with the load suspended from one end and a counter-weighted poise from the other. The poise was immersed in a tank of water. When a load was applied, the poise was lifted out of the tank, thus changing the height of the liquid. The weight was indicated by the height of water in a glass tube connected to the tank. Later, a modified design incorporated a chart and indicator which was activated by the movement of the beam.

From 1906, Avery favoured an 'aerostat' system for their self-indicating platform scales and weighbridges. This was the Stickig system, named after its inventor and patentee. The counter poise takes the form of two revolving drums with weights attached. As the load is applied to a central rod, the drums are rotated through steel ribbons. This causes the counter weights to swing outwards to a point of balance with the load. A linked quadrant and indicator needle allows the display of the weight on a circular dial.

History II:
The Industry of Precision

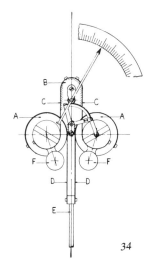

34

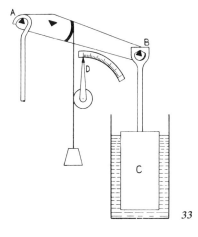

33

33   *Hydrostatic balance, introduced by Henry Pooley in 1874.*
     Indicator consists of a beam or lever, one end (A) of which is connected to a conventional load carrying system of a platform scale or weighbridge. From the other end (B) is suspended a counterpoise (C), dipping into a tank of liquid, usually water. When a load is applied on the weighbridge, the end (A) is pulled downwards and the counterpoise is raised out of the liquid, thereby losing part of its bouyancy until a position of balance is found automatically.

34   *'Aerostat' self-indicating mechanism.*
     Counterpoises take the form of two revolving drums (A) suspended from the frame (B) by steel ribbons (C). Drums are connected by another pair of ribbons (D) to the connecting rod (E). As the load is applied to this rod the drums tend to roll upwards and the attached weights (F) to swing outwards to a position of balance. A link and a quadrant transmit the movement of the drums to a spindle carrying an indicator, which is caused to travel round a circular dial.

History II:
The Industry of Precision

35 Dayton Royal computing scale by Henry Pooley and Son Limited.

36 Avery-Toledo cylinder scale with pendulum resistant.

The Stickig Aerostat was only one form of resistant used on circular dial self-indicating weighing machines. Another was the double-pendulum, developed by the Toledo Company, which worked in much the same way as the Aerostat but with certain advantages including a simpler indicating mechanism. Yet another was the Timson cam and lever resistant introduced by W. & T. Avery in 1928, forming the basis for all sizes of platform scales up to the very heaviest of weighbridges.

The Aerostat, double-pendulum, and the cam amd lever systems have been used on countless coin-operated scales, which were familiar sights outside chemists' shops and on railway stations until their numbers were diminished by the affluence of weight watchers who now consider that no bathroom is complete without personal scales—much to the scalemakers' delight!

History II:
The Industry of Precision

37

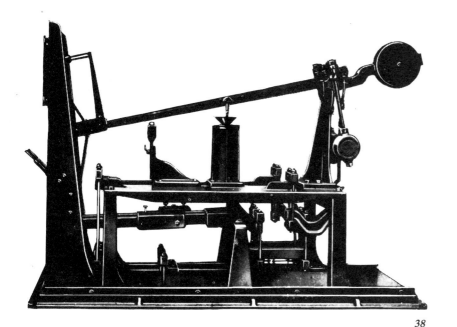

38

37 Avery semi-self-indicating. The major part of the load is balanced by the steelyard and a subsidiary pendulum resistant unit gives automatic indication of the minor divisions. c. 1924.

38 Mechanism of the Pooley quadrant indicator.

History II:
The Industry of Precision

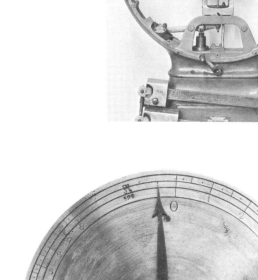

39 Interior mechanism of an Avery self-indicating scale. Late 1920s.

40 Toledo double-pendulum mechanism.

41 An indicating dial by Henry Pooley & Son Limited.

History II:
The Industry of Precision

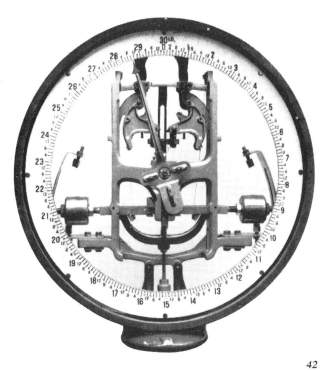

42

43

42 *Avery cam and lever resistant movement.*
43 *Avery double-pendulum mechanism.*

Examples of Early 19th Century Scientific Weighing Devices

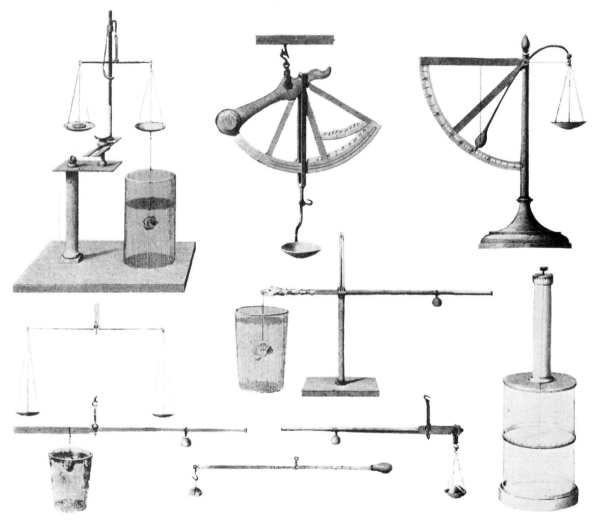

45 *Weights and measures—from a German edition of the* Orbis, *published 1833.*

TYPES OF SCALES AND TERMS USED

AEROSTAT. A mechanism developed in the early years of the 20th century by an inventor named Stickig, and used by W. & T. Avery Limited on self-indicating dial platform scales from 1906. Counterpoises are suspended revolving drums attached by steel ribbons to a connecting rod which is activated by the load platform. Under load, the drums roll upwards and attached weights swing outwards until a point of balance is attained. The mechanism is linked to an indicator needle showing the weight on a circular dial.

AGRICULTURAL WEIGHING MACHINES. Farmers' needs for weighing machines were, in the days of village blacksmiths, filled by locally-made steelyards and beam balances and almost invariably were of iron construction.

AUNCELL. An alternative name for the bismar.

AUTOMATIC SCALES. Mechanisms for continuous weighing of commodities such as grain. Used in docks, power stations, and food packing factories.

BANK SCALES. The classical Bank Scale is a beautifully fashioned and polished beam balance for weighing numbers of coins. Dial self-indicating machines are now replacing the beam balances.

BEAM SCALES. The original form of weighing device, evolved about 4000 BC. In essence, it comprises a beam suspended or supported at its centre. Pans, one for the load and the other for weights, are hung from the ends of the beam.

BERANGER LEVER SYSTEM. A standard arrangement of levers used on counter scales invented by Joseph Beranger in the mid-19th century. Each scale pan has a four-point support (see fig. 16).

BISMAR. An ancient form of counter-weighted lever balance. The load is suspended from one end of the lever and the counter-weight is fixed at the other. The fulcrum, usually a cord loop, is moved along the lever until the point of balance is found.

BOX-END PIVOT. Evolved in the 16th century as a means of achieving accuracy with beam balances (see fig. 20).

BRASS AND AGATE PIVOT. Patented by Sharkey in 1853 as a means of achieving greater accuracy with beam scales.

BUTTER SCALES. A specialised form of Roberval lever counter scales with the load pan substituted by a glazed ceramic circular plate.

CAM AND LEVER RESISTANT. A system for self-indicating dial scales, designed and patented by W. Timson and produced by W. & T. Avery Limited from 1928.

46 Beam balance for weighing sacks of corn. c. 1890.

47 Avery dial indicator.

CHEMISTS' SCALES. See Precision Scales.
COAL SCALES. Heavy-duty Roberval lever system scales, used by coal merchants in railway station yards in the past. The load pan has a folding frame to support the sack while being filled.
COIN SCALES. Small brass portable balances carried by merchants to test the weight of gold and silver coins at the time when 'clipping' and 'sweating' of metal from coins added a hazard to dealing. The scales are of 18th century and early 19th century origin and manufacture, and make very collectible subjects. Some are ingenious fold-away designs and all are contained in attractive wooden cases with partitions for weights. Not to be confused with bank scales which are intended for counting coins.
COMPUTING SCALES. Alternative name for Price-Indicating scales.
CONFECTIONERS' SCALES. Either beam balances or Beranger counter scales. The load pan, made from brass, is scoop-shaped to facilitate pouring sweets into paper bags.
CORD-PIVOT. The earliest method of suspending scale pans from beam balances. In the most primitive form the cord passes through a vertical hole in the beam end and is secured by a knot. Attempts were later made to increase accuracy by finding means of bringing the cord out of the extreme end of the beam (see fig. 20).
DANISH STEELYARD. Alternative name for the bismar.

Types of Scales and Terms Used

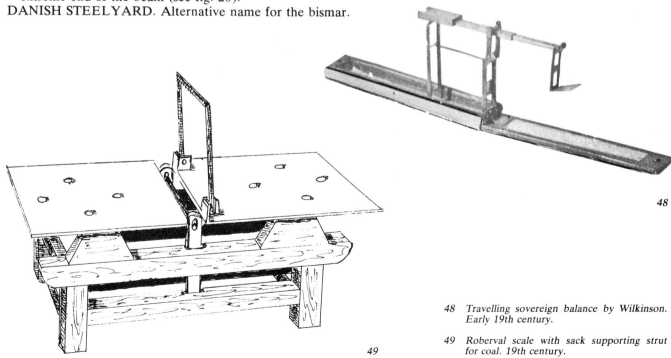

48 Travelling sovereign balance by Wilkinson. Early 19th century.

49 Roberval scale with sack supporting strut for coal. 19th century.

Types of Scales
and Terms Used

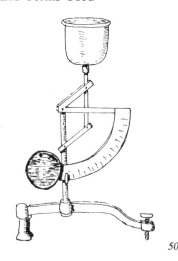

50

51

50 Quadrant balance for egg grading.

51 Semi-self-indicating 'fan' scale.

52 Weighing machine for colliery tubs.

DIAL RECORDING SCALES. Self-indicating scales also capable of printing the weight on a ticket or on tape.

DOUBLE-PENDULUM MECHANISM. A resistant mechanism developed by the Toledo Scale Company in the USA for self-indicating dial scales.

DUTCH-END PIVOT. Devised in the 18th century as a means of increasing accuracy of beam balances (see fig. 20).

EGG BALANCES. These are pendulum-type balances with indicating curved chart plates. An egg-cup took the place of the load pan carried at the top of a vertical stem.

ELLIPTICAL SPRINGS. See Spring Balances.

FAN SCALES. Self-indicating scales with the chart displayed at the head of a fan-shaped housing.

FULCRUM. The fixed point on which the suspended or supported beam or lever moves.

HELICAL SPRINGS. See Spring Balances.

HYDROSTATIC BALANCE. A mechanism introduced in 1874 by Henry Pooley for self-indicating heavy industrial scales. It comprises a beam with the counter weight immersed in water to provide the resistant and, in some cases, to indicate the weight of the load (see fig. 25).

IMPERIAL SCALES. A modification of the Roberval lever mechanism, in which the stays and legs are arranged above the scale pans and beam. (see fig. 12).

52

JOCKEY SCALES. A steelyard type of balance for personal weighing, with the subject sitting in a suspended chair. Popular with 'guess-your-weight' showmen at fair grounds.

KNIFE-EDGE. To reduce friction at pivots and fulcrum, one of the metal surfaces is reduced at the point of contact, thus increasing the sensitivity and accuracy. It comprises a steel or agate prism with its sharp edge resting on a polished plane.

LETTER SCALES. See Postal Balances.

LOTUS-ENDED BEAM. Developed by the ancient Egyptians in an attempt to increase the accuracy of beam balances (see fig. 20).

PACKET SCALES. See Postal Balances.

PENDULUM BALANCE. A balance in which the counter-weight poise or resistant follows the action of a pendulum.

PHANZELDER SCALE. A variation of the Beranger counter scale.

PLATFORM SCALES. A weighing machine with the load carried on a platform, as in the case of a weighbridge. Generally credited as the invention of John Wyatt about 1741, the platform is supported by a lever or compound levers linked to a steelyard or other resistant and weight-indicating device.

POISE AND SLIDING POISES. The counter-weights against which a load is measured. The sliding poise, as its name suggests, slides along a steelyard, thus indicating the weight by showing the graduation at the point of balance.

POSTAL BALANCES. Machines specially designed for assessing the weight of letters and packets. They are either Roberval lever, pendulum, or spring balance type. They usually carry the postage rate of the period in which they were made.

PRECISION BALANCES. Highly accurate beam balances for scientific use. Their essential ingredients are: delicately balanced knife-edges and planes, light beam, and a means of lifting the knife-edge at the fulcrum. Great advances have been made in the design of precision balances since the middle of the 18th century. The measurement of one part in half a million is possible with good precision balances.

PRICE-INDICATING SCALES (also known as Computing Scales). Developed in the last ten years of the 19th century by the American companies of Stimpson, Dayton, and Toledo.

PUNDLER. A wooden-armed steelyard. Some surviving examples are very large, exceeding 6ft in length and with stone poises weighing over 30lb.

QUINTENZ SCALES. A type of platform scale using a steelyard coupled to a single lever supporting the platform. The Quintenz was introduced about 1820 and achieved more popularity on the mainland of Europe than in Britain.

Types of Scales and Terms Used

53

53 Vibrating bench platform scale with offset pillar. Up to 1 cwt. Made by Fairbanks, USA. c. 1905.

Types of Scales and Terms Used

54 Jewellers' balance with drawer for carat weights. Mid-19th century.

RESISTANT. Counter-weight, spring, or other device against which a load is measured.

RING AND HOLE PIVOT. Means of suspending scale pans to a beam balance. A method favoured by the Romans (see fig. 20).

ROBERVAL LEVER SYSTEM. Standard arrangement of levers used on counter scales based on Gilles Personne de Roberval's 'Static Enigma' of 1669. The beam, legs and stay maintain a parallelogram, notwithstanding the position of the load and weight pans supported by the legs.

ROMAN STEELYARD. Most of the steelyards that have survived from the time of the Roman Empire are of bronze, although it is not unknown for them to be formed from wrought iron. For description see Steelyard.

SACK SCALES. An overall term covering all weighing machines designed for the measurement of goods contained in sacks—they may be of the Roberval lever type used when filling sacks of coal, or of the steelyard pattern for use in agriculture.

SCIENTIFIC SCALES. See Precision Balances.

SECTOR SPRING. See Spring Balances.

SELF-INDICATING SCALES. Types of balances on which the weight is displayed on a chart, dial, or revolving drum. The invention of the self-indicating scale is attributed to Leonardo da Vinci.

SPRING BALANCES. Types of scales which depend on a spring as the resistant. The springs can be helical (corkscrew) which are either compressed or stretched when a load is applied, sector, C-shaped (as in the case of the 'Mancur' balance), or elliptical (as designed by Siebe and Marriott). All Spring Balances are self-indicating.

STEELYARD. A balance consisting of a lever with unequal arms, in using which a weight, or counter poise, is moved along the graduated beam of the lever. The steelyard takes its name not from the metal used—early Roman instruments were of bronze—but from a Low German word meaning *sample yard*.

NOTE. The Danish Steelyard is not a true steelyard but a bismar, depending on a fixed counter poise and moveable fulcrum.

SWAN-NECK PIVOT. A design for balance beam ends to increase accuracy of measurement.

SWINGING-PAN CYLINDER SCALES. The name given to the first of the price-indicating scale marketed by the Dayton Computing Scale Company in the USA.

TRUMPET-END PIVOT. A beam-end, in shape resembling the ancient Egyptian Lotus-end beam, but with an inserted ring in place of knotted cord suspension of the scale pans.

WEIGHBRIDGES. Owing their invention to the passing of the Turnpike Act of 1741, the first compound-lever weighbridge was built and installed at Birmingham by John Wyatt. Wyatt's compound-lever principle has remained the basis for most platform scales and weighbridges.

EXAMPLES OF BEAM BALANCES

55

56

55 Travelling tea taster's sampling balance. A silver sixpence was often used in place of a weight. By William Williams & Sons, London.

56 Apothecary weights by De Grave, Short & Fanner, London.

Examples of Beam Balances

57 Brass and copper beam balance. Probably a Turkish customs scale. Date unknown.

Examples of
Beam Balances

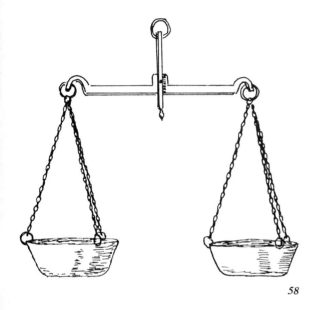

58

59

58 Tea balance for shop use.

59 Travelling sovereign beam balance. By Wilkinson. 19th century.

Examples of
Beam Balances

60   Box-end beam scale. 19th century.

Examples of
Beam Balances

61 Beam scale for agricultural use with swan-neck ends. The painted decoration is traditional. 19th century.

Examples of Beam Balances

62 English farmhouse beam scale. 17th century.

Examples of
Beam Balances

63 Precision balance. To weigh up to 2 oz.
By De Grave, Short and Fanner, London,
19th century.

Examples of
Beam Balances

64 Bank balance.

Examples of
Beam Balances

65 Beam balance with butter plate. 19th century.

Examples of
Beam Balances

66　*Avery precision balance.*

Examples of
Beam Balances

67 Dr Black's balance preserved in the New Chemical Department, University of Edinburgh. This is the balance used by the celebrated Joseph Black (1728–99) on which he performed some of his important experiments. The balance shows a very advance state of development for the latter part of the 18th century.

EXAMPLES OF STEELYARDS

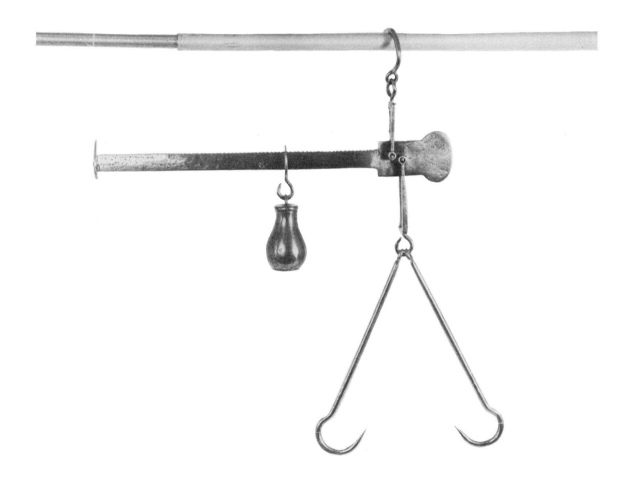

68 Steelyard of the 19th century.

Examples of
Steelyards

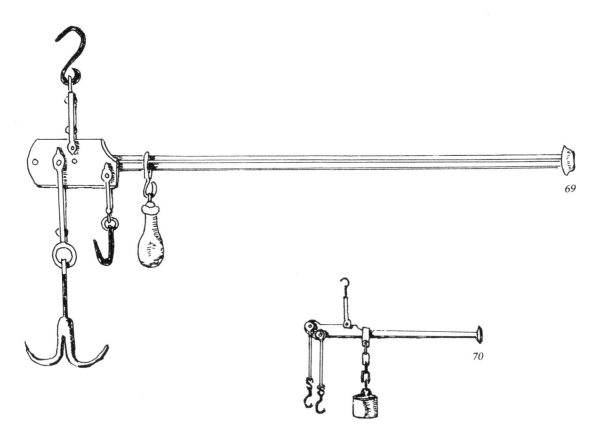

69 Butcher's steelyard. Mid-19th century.
70 Steelyard with lead-filled poise. Arabic. 18th century.

Examples of Steelyards

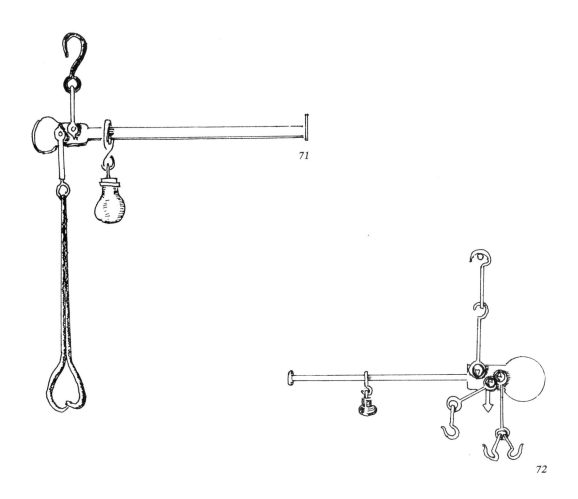

71 Steelyard. War Department issue. First World War.

72 Steelyard. Early 20th century.

Examples of
Steelyards

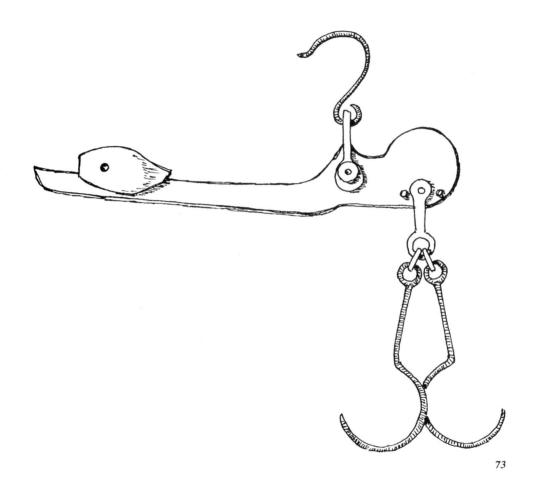

73

73 Baker's scale with flip-over poise for 2 lb and 4 lb loaves. Section 7 of the Bread Act, 1836, required bakers' roundsmen to carry scales. This act was repealed in 1926.

EXAMPLES OF SPRING BALANCES

75

74 *Spring balance with circular barrel. By George Salter. Late 18th century.*

75 *Two Salter pocket spring balances. Left: a circular barrel of the 19th century. Right: a Class II balance of the early 20th century.*

Examples of
Spring Balances

76 An early Salter spring balance in the Avery Historical Museum.

Examples of
Spring Balances

77 Sector spring balance.

78 Mancur spring balance.

Examples of
Spring Balances

79 Siebe spring balance.
80 Elliptical spring balance.

Examples of Spring Balances

81 Quadrant spring balance. Graduated to 200 lb. By George Salter. 19th century.

82 Agricultural spring balance with circular dial. By George Salter. 19th century.

EXAMPLES OF COUNTER SCALES

83   Panel scale. Dated 1912.

Examples of Counter Scales

84 Beranger counter scale. By Day & Millward, Birmingham. Late 19th century.

Examples of
Counter Scales

85 Brass counter scales. Early 20th century.

Examples of
Counter Scales

86 Beranger counter scales. To weigh up to 2 lb. By W. & T. Avery. Late 19th century.

Examples of
Counter Scales

KITCHEN SCALES

87

88

87 and 88   Two examples of Roberval kitchen scales.

SWEET SCALES

Examples of
Counter Scales

89

90

89 Roberval counter scales.

90 Counter scales marked 'French Machine'.
    Late 19th century.

EXAMPLES OF COIN BALANCES

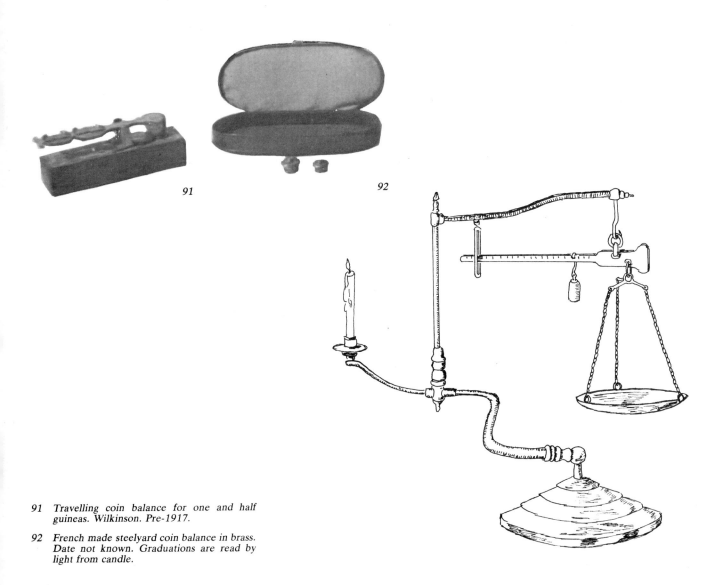

91 Travelling coin balance for one and half guineas. Wilkinson. Pre-1917.

92 French made steelyard coin balance in brass. Date not known. Graduations are read by light from candle.

Examples of
Coin Balances

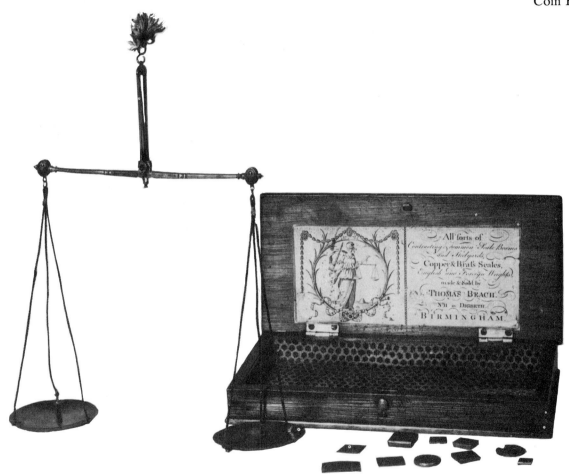

93   *Coin scale by Thomas Beach.*

Examples of
Coin Balances

94 *Wilkinson coin scale. c. 1780. Compare with
Fig. 92 (page 64).*

EXAMPLES OF LETTER SCALES

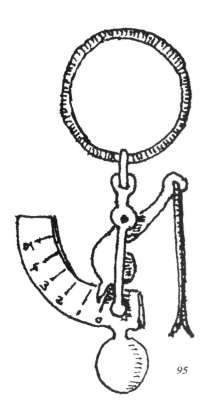

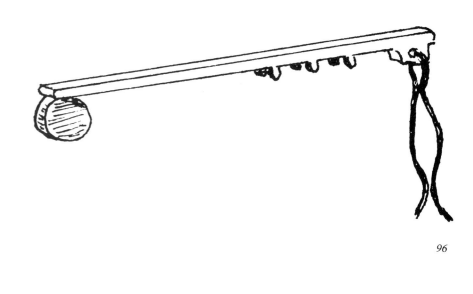

95 Quadrant letter balance. Probably French. Late 19th century.

96 A simple beam scale for weighing letters. The letter is placed in the clip and the beam balanced on a finger. To weigh 1 or 2 ounces, depending on the position of the finger.

Examples of
Letter Scales

97 Pocket pendulum balance.

Examples of
Letter Scales

98. 99, 100, 101 and 102   Examples of German-made letter balances. c. 1912.

Examples of
Letter Scales

103 Compression spring letter balance. Marked 1840.

104 Steelyard letter balance. 1880s.

105 Quadrant letter balance. Early 20th century.

Examples of
Letter Scales

106

107

108

106 Letter balance of 1912. Similar models were made by a number of scale makers from the late 19th century.

107 Quadrant letter balance. c. 1925.

108 Compression spring parcel balance. 19th century.

Evolution of the Pivot

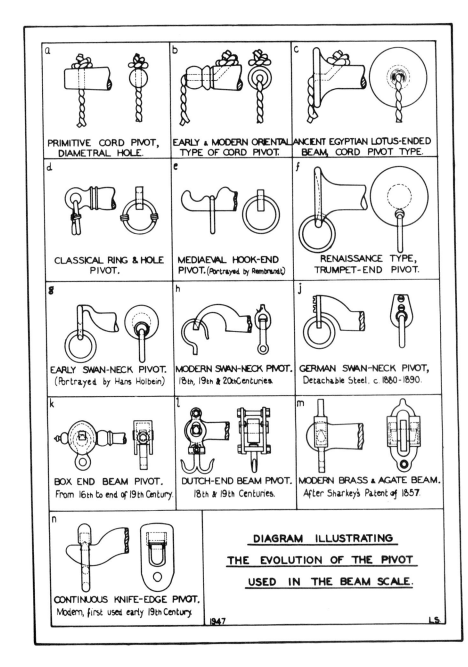

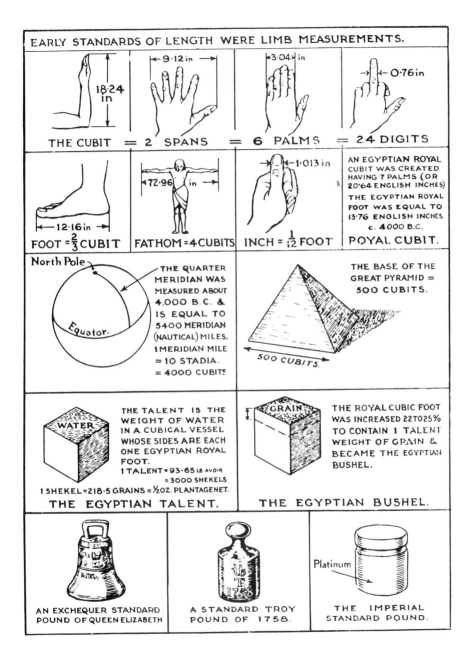

Weights and Measures

WEIGHTS AND MEASURES

111

112

111 A hentha-shaped weight from Burma.

112 Pyramid case for weights. Marked 1854.

The earliest standards of weight and measurement were based on human limb measurements and common objects: the digit (the breadth of the middle finger), palm (breadth of 4 fingers) equal to 4 digits, span (the distance between thumb and little finger when the hand is spread) equal to 3 palms, cubit (length of the bent forearm from the point of the elbow to the tip of the fingers). In 1101, Henry I commanded that the length of his stretched arm should define a yard.

Weights bore a relationship to standards of length: the talent was the weight of a cubic Egyptian Royal Foot (equal to 13·76in)—93·65lb. This was divided into 3,000 shekels (218·5 grains each) which came to be half the English ounce in Plantagenet times.

The first weights to be used on beam scales were rounded stones. Later they were cast in bronze, often in the shape of animals. The Egyptians favoured cows for the shape of their weights.

Animal shapes—either factual or mythical—were used for weights in parts of Asia until well into the 20th century. For instance, the Burmese used weights cast in the shape of the *hentha* bird which, legend held, found a home in Pegu (founded 573 AD) in Lower Burma.

The libra of 12 ounces (each of 2 shekels) was brought to Britain by the Romans. It was later reassessed and divided into 16 ounces.

This 16 ounce pound became known as the Tower Pound under the Normans—the mint being located at the Tower of London. From this pound were minted 240 silver pennies, giving the basis for British coinage until decimalisation.

Even so, formality did not always rule the coinage. In 1266, it was legally declared 'that an English *peny* called a sterling, round and without clipping, should weigh 32 wheat corns, from the midst of the ear'.

The Troy (Troie) pound originated at the Troyes Fair, one of the great continental markets of the Middle Ages, an by an Act of 1527, 'the pound Towre shall be no more used, but all manner of gold and silver shall be weighed by the pound Troye which exceedeth the Pound Towre in weight 3 quarters of the ounce'.

However, the old Tower Pound did not die and, renamed the Imperial Pound, was reinstated by Elizabeth, and the Troy pound retained only for bullion.

Physically, the present Imperial Standard Pound was made in 1844. It is a cylinder of platinum with a groove for an ivory lifting fork.

Weights and Measures

113

114

113 Porcelain weights. Pre-1917. The bar weight is marked 'Day & Millward, Aldgate E.C. Manchester. Quotations given for yearly maintenance. Machines lent on hire for stocktaking'.

114 Bell and bucket weights. Late 18th and early 19th centuries.

Weights and Measures

Some Units of Weight Established by Custom

Bag of Coffee: 112 to 180lb according to custom.
Bag of Cocoa or Sago: 112lb.
Bag of Sugar: 224lb.
Bale of Cotton: 400 to 500lb (American).
              700 to 740lb (Egyptian).
              500 to 600lb (Indian).
Bar of Gold (Mint): 400oz Troy.
Bar of Silver (Mint): 1,000 to 1,100oz Troy.
Barrel of Beef: 200lb.
Barrel of Butter: 106 or 256lb.
Firkin of Butter: 56lb.
Yard of Butter: 1lb.
Barrel of Flour: 196 or 228lb.
Peck of Flour: 14lb.
Sack of Flour: 280lb.
Barrel of Gunpowder: 100lb.
Barrel of Raisins: 112lb.
Barrel of Soft Soap: 256lb.
Firkin of Soap: 64lb.
Bolt of Canvas: 35lb.
Bushel of Barley: 47lb.
Bushel of Oats: 38lb.
Bushel of Wheat: 60lb.
Truss of Hay (new): 60lb.
Truss of Hay (old): 56lb.
Truss of Straw: 36lb.
Chest of Tea: 80 to 84lb.
Load of Bricks: 500lb.
Pig of Ballast: 56lb.
Hogshead of Tobacco: 12 to 18cwt.
Pack of Wool or Flax: 240 or 480lb.
Sack of Wool: 364lb.
Pocket of Hops: 168 to 224lb.
Seam of Glass: 120lb.

Weights and Measures

115 Trial of weights and measures under Henry VII.

WEIGHTS AND MEASURES IN 1826

Extract from *The Every-Day Book* by William Hone, 1827

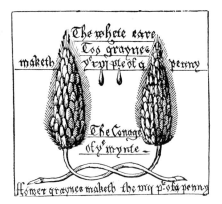

The alteration of the standard this year, in order to enforce its uniformity throughout the kingdom, however inconvenient to individuals in its first application, will be ultimately of the highest public advantage. The difference between beer, wine, corn, and coal measure, and the difference of measures of the same denomination in different counties, were occasions of fraud and grievance without remedy until the present act of parliament commenced to operate. In the twelfth year of Henry VII, a standard was established, and the table was kept in the treasury of the king's exchequer, with drawings on it, commemorative of the regulations, and illustrating its principles. The original document passed into the collection of the liberal Harley, earl of Oxford, and there being a print of it with some of its pictorial representations, an engraving is here given of the mode of trial which it exhibits as having been used in the exchequer at that period.

From the same instrument is also taken the smaller diagram. They are curious specimens of the care used by our ancestors to establish and exemplify rules by which all purchases and sales were to be effected. In that view only they are introduced here. Conformity to show the new standard is every man's business and interest, and daily experience will prove it wisdom and justice. It would be obviously inexpedient to state any of the parliamentary provisions in this work, which now merely records one of the most remarkable and laudable acts in the history of our legislation.

117 The Gold Weigher, by Van Ryyn Paul Rembrant (1606–69)

118 The Balance Maker c. 1695.

DIRECTORY OF MARQUES, INVENTORS AND MANUFACTURERS

It cannot be claimed that the following list is a comprehensive survey of all the names connected with the long and complex history of weighing machines. Hitherto unknown makes are continually coming to light and even while this book is being prepared, additions and corrections are being made. However, the Directory will be useful to collectors when tracing histories of scales and balances.

ABRAHAM & COMPANY, Glasgow. Precision instrument and scale maker.

ADIE, London. Precision instrument and scale maker. Active at least c. 1860.

ALMOND, William. 281 Borough High Street, London. Active up to early 1820s.

ANDERSON, Js. 19 Wardour Street, London. Active at least in the 1820s.

AUTOMATIC SCALE COMPANY LIMITED, London and Manchester.

AVERINK. Amsterdam, Holland. Precision instrument and scale maker. Active at least c. 1750.

AVERY, W. Redditch. Makers of postal balances.

AVERY LIMITED, William & Thomas, Soho Foundry, Birmingham. At Digbeth at least in the 1820s.

BAIRD & SON, Glasgow. Precision instrument and scale maker.

BANFIELD, S. Ship Street, Brighton. Also at Hastings. Active from the mid-19th century until c. 1965.

BARTLETT, J. Long Lane, West Smithfield, London. Established c. 1825.

BASSINGHAM, S. 63 Bethnal Green Road, London. Established c. 1825.

BASTICK & DRIVER, 2 Holywell Row, London. Established c. 1825.

BATHER, George. 48 Great Windmill Street, Haymarket, London. Active at least in the 1820s.

BATHER, J. Chad's Row, Gray's Inn Road, London. Established c. 1825.

BEACH, Thomas. Digbeth, Birmingham. Produced coin weighing balances up to the early 19th century.

BERANGER, Joseph. A 19th century French inventor who devised a type of balance that bears his name. A very accurate scale but more complicated than the Roberval design.

BERKEL AUTO SCALE COMPANY LIMITED, London.

Directory of Marques, Inventors and Manufacturers

BERRY & MACKAY. Precision instrument and scale maker.
BITHRAY. Royal Exchange, London. Maker of precision instruments and scales. Active at least c. 1830.
BLACKBURN, John. 126 Minories, London. Active at least in the 1820s.
BOGAN, C. H. Grimsby. Maker of precision instruments and scales. Active at least c. 1850.
BOURNE & PAINTER, Broad Street, Birmingham. Active at least in the 1820s.
BOURNE & WRIGHT, Great Charles Street, Birmingham. Active at least in the 1820s.
BRIGNALL, J. A. 16 George Street, Hastings. Currently active.
BROWN & SHARPE MFG COMPANY, Providence, Rhode Island, USA. Active at least in the 1930s.
BURLEY. Frome. Maker of precision instruments and scales. Active at least c. 1725.
BUTTERFIELD, Paris, France. Maker of precision instruments and scales. Active at least c. 1750.
CAMERON, Dundee. Precision instrument and scale maker.
CARLOW, David. Glasgow. Mid-19th century maker of marine models, precision instruments and scales. The company survives as Carlow Engineering.
CARY. London. Maker of precision instruments and scales. Active at least c. 1780 until c. 1910.
CHISSHOLME, James. 15 Great Poultney Street, Golden Square, London. Active up to the mid-1820s.
CLAIS, John. London. Granted British patent on April 30, 1772, covering an indicating-dial weighing machine.
CLARK, John. 38 Cow Cross, West Smithfield, London. Active at at least in the 1820s.
CLARK, William. 29 Great Sutton Street, Clerkenwell, London. Active at least in the 1820s.
COOKSEY & COMPANY, Samuel. West Bromwich. Makers of a patent steelyard. Active in the 1820s.
COPE, Charles. John Street, Birmingham. Scale maker. Active at least in the 1820s.
CORRADI, Zurich, Switzerland. Precision instrument and scale maker.
DAVIS, J. Edinburgh. Precision instrument and scale maker.
DAY & MILLWARD, Birmingham. Products included Berenger counter scales. Active at least in the late 19th century.

DAYTON COMPUTING SCALE COMPANY, Dayton, Ohio, USA. Makers of automatic price-indicating scales from *c.* 1900. By arrangement, production in Britain was undertaken by Henry Pooley & Son Limited, when it became unofficially known as the 'swinging pan cylinder' scale.

DE GRAVE & SON, Mary. 50 St. Martins-le-Grand, London. Active at least in the 1820s.

DE GRAVE, SHORT & FANNER, London. Active at least in the early 19th century. Later the name was changed to DE GRAVE, SHORT & COMPANY LIMITED, whose address was 102 Naylor Road, London, S.E.15.

DENISON, England. Maker of suspended weighing machines in the 1930s.

DOWNIE, T. Hamburg, Germany. Maker of precision instruments and scales. Active at least *c.* 1850.

DOYLE, John. St. Thomas's, The Borough, London. Makers of steelyards. Established *c.* 1825.

DUBOIS. Precision instrument and scale maker. Active at least *c.* 1760.

DUNGATE & SONS, W. 132 Landor Road, London, SW9.

DUTTON & SMITH, 248 Tooley Street, London, SE. Active at least in the 1820s.

EAYRE. The surname of an engineer connected with the development of platform weighing machines.

EDWARDS, W. Glasgow. Precision instrument and scale maker.

ELLAW'S WEIGH-O-PHONE. A speak-your-weight machine introduced in 1929.

ELY, Newport. 28 Western Road, St. Leonards-on-Sea, Sussex. Active *c.* 1895 until *c.* 1937.

EVERITT, Percy. Granted British patent on December 13, 1894, covering a coin-operated weighing machine.

FAGE, William. 38 Brown's Lane, Spitalfields, London. Active at least in the 1820s.

FAIRBANKS, THADDEUS & ERASTUS. American inventors who were granted a US patent in 1831, covering a compound lever scale. A British patent granted in 1833 was declared invalid when scale makers in this country proved that a similar invention had been in use for a number of years. Henry Pooley later produced the Fairbanks platform scale which was first used on the Manchester-Liverpool Railway in 1835.

Directory of Marques, Inventors and Manufacturers

Directory of Marques, Inventors and Manufacturers

FORTIN, Paris. Maker of precision balances. Active in the late 18th century.
FOURCHE, Paris. Maker of precision balances. Active in the 1780s.
FLETCHER, William. Princes Street, Mile End, London. Established c. 1825.
GARLAND & BRIDGMAN, 48 Ball Street, Birmingham. Active at least in the 1820s.
GERRARD, W. South Castle Street, London. Maker of precision instruments and scales. Active at least c. 1880.
GREENOUGH, Thomas. Maker of precision instruments and scales. Active at least c. 1755.
GREGORY, H. London. Maker of precision instruments and scales. Active at least c. 1730.
GROUT, John. 35 Fashion Street, Spitalfields, London. Maker of beam scales. Active at least in the 1820s.
HIATT, Henry. 83 Old Street Road, London. Scale maker. Active at least in the 1820s.
HIATT, Mary A. 18 Vine Street, London. Established c. 1825.
HAWKES, Joshua. 38 Princes Street, Whitechapel, London. Established in the 1820s.
HEARD, Thomas. 77 Long Lane, West Smithfield, London. Established in the 1820s.
HEARD JUNIOR, Thomas. 39 St. John Street, Clerkenwell, London. Established in the 1820s.
HENN, William. Lancaster Street, Birmingham. Active at least in the 1820s.
HERBERT & SONS. Smithfield, London. Established in the 19th century. Still active.
HOPPE. London. Maker of precision instruments and scales. Active at least c. 1800.
HUGHES, England. Maker of spring balances. Active at least in the 1930s.
JONES, Thomas. London. Maker of precision instruments and scales. Active at least c. 1820 until c. 1930.
KENDALL, S. Park Street, Birmingham. Maker of steelyards. Active at least in the 1820s.
KRUPS, Germany. Made kitchen spring balances in the early 20th century.
LEAKE, George. 15 Great Poultney Street, London. Established c. 1825.
LEONARDO DA VINCI (1454–1519). Credited with the design of the first recorded self-indicating scale. Principle was not adopted by scale makers until the 19th century.

MacCRAIGHT, J. C. 88 Goswell Road, London. Active at least in the 1820's.

MARILE, Paris. Maker of coin operated scales. Active in the early 20th century.

MARRIOTT, H. Worked with Augustus Siebe on the development of an elliptical flat strip steel spring balance, which was made by Henry Pooley & Sons from 1853.

MARRIOTT, W. Probably a son of H. Marriott. It was W. Marriott's signature that appeared on an agreement with Henry Pooley & Sons in 1853, under which the elliptical flat spring balance was produced.

MEDHURST, George. 1 Denmark Street, Soho, London. Active at least in the 1820's.

MEGNIE, Paris. Maker of precision balances. Active in the 1780s.

MESSER & COMPANY, G. B. London. Maker of precision instruments and scales. Active at least *c.* 1870.

MEYMOTT, S. 64 Bishopsgate Within, London. Active at least in the 1820s.

MILLIOT, Rue de Cardinal. Paris. Makers of counter scales.

MORDAN, S. London. Maker of scales in the 19th and early 20th centuries, including postal balances.

MORVAN, C. & B. Birmingham. Makers of platform scales.

NAIRNE. London. Maker of precision instruments and scales. Active at least between *c.* 1760 and *c.* 1800.

NEWMAN, S. 27 Crispin Street, Spitalfields, London. Active up to the mid-1820s.

NEWMAN, T. P. 27 Union Street, Bishopsgate, London. Established *c.* 1825.

NICHOLL, William Lewis. 166 and 167 Aldersgate Street, London. Active at least in the 1820s.

NOVINGTON, William Michael. 20 Fann Street, Goswell Street, London. Active up to the mid-1820s.

PALLET, Elizabeth. 91 Leadenhall Street, London. Active at least in the 1820s.

PARNALL & SONS LTD, Narrow Lane, Bristol. Makers of all kinds of balances. London Depot: 10 Rood Lane, Eastcheap.

PAYNE, Benjamin. 395 Strand, London. Active at least in the 1820s.

PEARCE, Samuel. 46 Fuller Street, Shoreditch, London. Established *c.* 1825.

PECK, John. 98 Great Guildford Street, Southwark, London. Active at least in the 1820s.

PHILLIPS, Samuel. 5 Coal Yard, Drury Lane, London. Established *c.* 1825.

Directory of Marques, Inventors and Manufacturers

Directory of Marques, Inventors and Manufacturers

POOLEY & SON LIMITED, Henry. Liverpool. Undertook the manufacture in Britain of the Fairbanks compound lever platform scale, used first on the Manchester–Liverpool Railway in June 1835. From 1853, Pooley fitted spring dials to the Fairbanks pattern platform scales. The dial mechanism was based on the spring balance design developed by Siebe and Marriott. Henry Pooley introduced hydrostatic balances in 1874 which were appplied to weighbridges. Quadrant indicator scales for heavy industrial use were made by Pooley from 1907. Pooley's London office in 1864 was at 89 Fleet Street.

PORTER, S. Active at least c. 1820. Some of Porter's products carry the legend 'Patronized by the King'.

POUPARD, Abraham. 3 Lower Chapman Street, London. Established c. 1825.

RIPLEY, Thomas. Hermitage Bridge, London. Maker of precision instruments and scales. Active at least c. 1770.

ROBERVAL, Gilles Personne de. In 1669 de Roberval invented the 'Static Enigma', the principle of which puzzled the scientific minds of the period. The principle was developed for use on counter scales. In one variation, the Imperial, the stays are arranged above the beam.

ROBINSON, John. 5 Bethnal Green Road, London. Established c. 1825.

ROBINSON, Thomas. Mill Lane, Dibeth, Birmingham. Active at least in the 1820s.

ROSENBERG, Carl. Operated the first recorded public weighing machine in 1799 at Bath.

RUBIDGE, Charlotte. 18 Great Eastcheap, Fish Street Hill, London. Active up to the mid-1820s.

RUST, Richard. Precision instrument and scale maker. Active at least c. 1760.

SAFAA, (Societe Anonyme Francaise des Appareils Automatiques), 75 Rue la Condamine, Paris. Makers of coin operated weighing machines since 1885.

SALTER & COMPANY, George. West Bromwich. Scale makers since 1760. An early member of the family, Richard Salter, produced spring balances since c. 1770. In 1901 a company advertisement read: 'Salter's, the originators and inventors of the spring balance in its original forms, not only maintain the lead, but through the excellence and moderate price of their work are practically the only British makers of these goods. The compactness, handiness, and portability of the balances render them the greatest service for Military and Naval use, and the world-wide demand for them (with markings available for various countries, from metric system to Chinese) attests their utility and worth.'

The Salter's Auto-Phonograph, a speak-your-weight machine, was manufactured from 1929.

SAWGOOD, S. 6 Hoxton Market, London. Established *c.* 1825.
SHARKEY. The surname of the patentee of the brass and agate beam pivot. Patent granted in 1853.
SHERIDIN. Precision instrument and scale maker. Active at least *c.* 1790.
SHURY, Samuel. St. Mary Axe, London. Between 1823 and 1826 two further addresses appear in the London trade directories: Lambs Passage, and Artillery Gardens.
SIEBE, Augustus. Designer, in the early 19th century, of a balance using flat strip steel springs. Later, with H. Marriott, Siebe developed a true elliptical flat strip steel spring balance which was made by Henry Pooley from 1853.
SILBERRED. London. Precision instrument and scale maker. Active at least *c.* 1800.
SMITH, Thomas. 54 The Minories, London. Established *c.* 1825.
SPENCER, BROWNING & COMPANY LIMITED. London. Makers of precision instruments and scales between *c.* 1830 and *c.* 1890.
STALKER, D. Leith, Scotland. Precision instrument and scale maker. Active at least *c.* 1840.
STICKIG. Surname of a weighing mechanism known as the Aerostat. The system was used on W. & T. Avery platform scales and weighbridges from 1906.
STIMPSON COMPANY, Detroit, Mich. Maker of price-indicating scales from 1897. Sold mainly in the USA.
SYKES, C. McG. Inventor of a semi-self-indicating platform scale made by W. & T. Avery from 1924.
TANGYE. Makers of scales and measuring instruments.
TAYLOR & COMPANY, Mrs. Janet. The Minories, London. Makers of precision instruments and scales. Active at least *c.* 1830.
THOUGHTON (or Troughton). London. Maker of precision instruments and scales. Active at least *c.* 1820.
THOUGHTON & SIMMS. London. Makers of precision instruments and scales. Established 1826 and active at least until *c.* 1925.
THROPP & SON, William. 26 Little Street, Seven Dials, London. Patentees. Established *c.* 1825.
TIMSON, W. Patentee of a cam and lever resistant mechanism made by W. & T. Avery Limited from 1928.
TOLEDO SCALE COMPANY. USA. Makers of automatic price-indicating cylinder scales, using improvements to De Vilbiss' patents. Made in England by W. & T. Avery Limited as the Avery-Toledo cylinder scale.
TOMBS, Benjamin. 148 Goswell Street, London. Active up to mid-1820s.
TOOMBS, Thomas. 29 Church Street, Spitalfields, London. Established *c.* 1825.

Directory of Marques, Inventors and Manufacturers

TOWER & COMPANY, J. W. Head Office at Widnes with other addresses at Manchester and Liverpool.

TRIPODI, Joseph. Genoa, Italy. Invented and installed the first known speak-your-weight machine in 1928 at Genoa.

TURNBULL & COMPANY. Edinburgh. Precision instrument and scale makers.

URINGS, I. London. Maker of precision instruments and scales. Active at least c. 1750.

VANDOME, Richard. 117 Leadenhall Street, London. Active at least in the 1820s.

VILBISS, De. Patentee of a fan computing scale with pendulum resistant.

WALKER, Edward. 19 Wych Street, Strand, London. Established c. 1825.

WENBURN, Robert. 190 High Holborn, London. Active at least in the 1820s.

WHITBREAD, G. London. Maker of precision instruments and scales. Active at least c. 1860.

WHITE, Thomas. London. Maker of precision instruments and scales. Active at least c. 1740.

WHITFIELD & SON, Edward. Church Street, Birmingham. Active at least in the 1820s.

WHITMORE & SON, W. Newhall Street, Birmingham. Active at least in the 1820s.

WILKINSON. Scale maker. Produced coin weighing balances in the 18th century.

WILLIAMS, Thomas. 71 Cannon Street and 4 Abchurch Yard, London. Active at least in the 1820s.

WILLIAMS & SONS, William. London EC. Maker of scales, including travelling tea tasters' sampling balances.

WINFIELD, W. R. Makers of postal spring balances. A surviving example is dated 1840.

WOOD, Robert. 7 West Smithfield and 15 Queen Street, Cheapside, London. Active at least in the 1820s.

WORTHINGTON. Maker of precision instruments and scales. Active at least c. 1860.

WYATT, John. Birmingham, England. Employed by Matthew Boulton at the Soho Manufactury, Birmingham. Built and installed several early weighbridges including, in 1741, the first of the lever-balanced machines at Birmingham.

YEOMANS. A name connected with the development of platform weighing machines.

YOUNG, George. 105 London Road, Birmingham. Established c. 1825.

YOUNG & SON, John. 5 Bear Street, Leicester Street, London. Active at least in the 1820s.

119 Miniature Roberval counter scale. Made 1885 by Newport Ely when apprenticed.

PYKNOMETERS AND HYDROMETERS

Sikes's Hydrometer.
A, Weight to be slipped on at c.

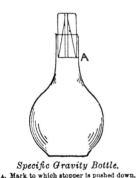

Specific Gravity Bottle.
A, Mark to which stopper is pushed down.

Specific Gravity is the comparison of the heaviness of a substance with that of a standard substance, and may thus be defined as the number of times the weight of a certain volume of the substance contains the weight of the same volume of the standard. In the case of solids and liquids, water at the point of its maximum density—at 4 degrees Centigrade—is usually taken as the standard.

For solids and liquids two principles are used to determine specific gravities:
i. by weighing measured volumes.
ii. by measuring buoyancy, and hydrostatic pressure.

In obtaining a comparative weight of a given volume of solids the difficulty is that they cannot be made to fit exactly into the measuring vessel. This is overcome by filling the empty spaces with water, the weight of which is subtracted from the weight of the water filling the whole vessel, thus obtaining the weight of the water equal in volume to the substance. The vessels used are called pyknometers or specific gravity bottles—essentially a bottle with a perforated stopper which can be filled to the same level each time. The buoyancy methods of determining specific gravities depend on the principle discovered by Archimedes, that a body immersed in a liquid is buoyed up by a force equal to the weight of liquid it displaces. Thus, if an object is weighed first in air, and then while it is suspended in water, it weighs less the second time by an amount equal to the weight of water in volume to itself.

This weight, divided into the weight of the object, gives the specific gravity. In the case of liquids it is only necessary to find the loss of weight of any object both in the liquid and in water. This can be done by using a hydrometer.

Hydrometers are of two types: of fixed and variable immersion.

Nicholson's Hydrometer is of the first kind, and comprises a hollow brass cylinder with conical ends, a weighed basket is fixed at the lower end to force it to float upright, and a pan supported by a vertical strut at the upper end. The instrument is adjusted by weights on the pan so as always to be immersed to a marked point on the strut. The difference in weights required to produce this, when immersed in water and then in the liquid to be measured, will give the specific gravity.

The specific gravity of a solid can also be found if it is first placed in the upper pan and then in the basket, the instrument being adjusted with weights each time.

Mohr's specific gravity balance is on the same principle as Nicholson's hydrometer, a plummet being sunk to the same point by placing rider-form weights on a balanced beam, from which the plummet is suspended. Mohr's instrument is easier to use and specific gravities can be found without calculation.

Specific gravity beads are fixed immersion hydrometers, consisting of small differently weighted bulbs that will either float or sink in a liquid according to its density. The bead that floats to a mark on the bulb gives the specific gravity of the liquid.

Hydrometers of variable immersion are light hollow spindles weighted by mercury or lead shot. Marked graduations are unequal, for as the volumes immersed are inversely indicative of the densities of the liquids, the spaces showing equal increments in density also diminish.

Variable immersion hydrometers are made of special forms and sizes to test particular liquids: alcohol, milk, etc. The most important is Sike's hydrometer, which is standard equipment for finding the strength of alcohol.

Sike's hydrometers are attractive instruments and at least one should be included in a collection of scales and balances.

Pyknometers and Hydrometers

121 Sike's hydrometer.

122 Pendulum scale (no springs) for light and heavy weighing. Up to 10 kgs. Made in Sweden. 19th century.

CITY OF BIRMINGHAM MUSEUM

A collection of historic exhibits, which formerly belonged to The Weights and Measures Department of the City of Birmingham, is displayed at the time of publication of this book, in the Local History Galleries of the City of Birmingham Museum.

Some 100 items make up the collection, which includes the following:

Model of Wyatt's Weighbridge, 1741.
Model of Improved Weighbridge, 1793.
Model Cart Weighing Steelyard.
(These three models were built in the first half of the 20th century by Mr. A. G. Birch, one time employee of the City of Birmingham Weights and Measures Department.)
Photograph: *Old Steelyard Cart Weighing Machine* at Woodbridge, Suffolk.
Weight Ticket issued at Wyatt's first Weighbridge, 1772.
A pre-1888 'Borough of Birmingham' Beamscale.
A post-1889 'City of Birmingham' Beamscale.
A Beamscale, probably from the 17th century.
19th century Butter Scales.
Early 19th century Coin Scales.
Japanese Dispensing Scale.
18th century Bismar.
Engraving: *Fish Woman of Stockholm,* with Bismar, 1882.
Corn Balance, or Chondrometer of the early 19th century.
Model of a Sack Weighing Deadweight Machine, 19th century.
Mancur, or C Spring Balance, *c.* 1880.
Bushel Measure, 1674—'Manor of Birmingham'.

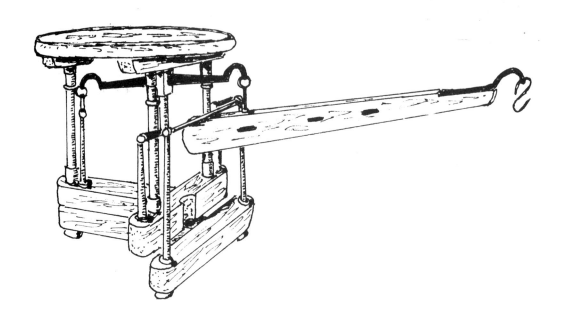

123 An unusual weighing machine thought to be c. 1760. Purpose unknown, possibly for weighing children.

ACKNOWLEDGMENTS

The author and publishers wish to express their gratitude for the kind co-operation of the following: Mr J. A. Brignall of Brignall's Scales, Hastings; Mr David Hance; Mr Stephen R. Hargreaves, West Midlands Metropolitan County Council; Mr Stephen J. Price, Department of Archaeology, City of Birmingham Museums; Mr A. Whittle, General Secretary of the National Federation of Scale and Weighing Machine Manufacturers; Mr A. R. Penzer of W. & T. Avery Ltd., who supplied many of the illustrations; Miss Carol Edwards of Granny Goods, East Molesey and Mrs E. M. Spark of Cobham.

BIBLIOGRAPHY

Buchanan, TABLES OF WEIGHTS AND MEASURES, 1838.
Chisholm, H. W. WEIGHING AND MEASURING, 1873.
Sanders, L. A SHORT HISTORY OF WEIGHING, 1947.

124 and *125* Coin operated scales by SAFAA of Paris.

NOR:RD

**NORFOLK
COUNTY
LIBRARY**

**REFERENCE
DEPARTMENT**

R681:2

This book
may not be taken away